Samuel Finley Breese Morse

Samuel Morse Petition for Extension of Patent

Samuel Finley Breese Morse

Samuel Morse Petition for Extension of Patent

ISBN/EAN: 9783744799379

Printed in Europe, USA, Canada, Australia, Japan

Cover: Foto ©berggeist007 / pixelio.de

More available books at **www.hansebooks.com**

A

Before the Commissioner of Patents.

IN THE MATTER OF THE APPLICATION

OF

SAMUEL F. B. MORSE,

FOR

AN EXTENSION FOR SEVEN YEARS OF LETTERS PATENT GRANTED TO
HIM JUNE 20th, 1840, REISSUED JANUARY 15th, 1846, AND
AGAIN REISSUED JUNE 13th, 1848.

FOR THE

ELECTRO-MAGNETIC RECORDING TELEGRAPH.

ARGUMENT IN FAVOR OF THE EXTENSION.

WASHINGTON:
PRINTED BY JOHN T. AND LEM. TOWERS.
1854.

Before the Commissioner of Patents.

IN THE MATTER OF THE APPLICATION

OF

SAMUEL F. B. MORSE,

FOR

AN EXTENSION FOR SEVEN YEARS OF LETTERS PATENT GRANTED TO HIM JUNE 20th, 1840, REISSUED JANUARY 15th, 1846, AND AGAIN REISSUED JUNE 13th, 1848,

FOR THE

ELECTRO-MAGNETIC RECORDING TELEGRAPH.

ARGUMENT IN FAVOR OF THE EXTENSION.

WASHINGTON:
PRINTED BY JOHN T. AND LEM. TOWERS.
1854.

In the matter of the application of Samuel F. B. Morse, for an Extension of his letters Patent for an Improvement in the mode of communicating Information by Signals, by the application of Electro-Magnetism, granted June 20, 1840, as reissued January 20, 1846, and again reissued June 13, 1848.

It is respectfully submitted on behalf of the said applicant that the provisions of law and the regulations of the Patent Office have been complied with, and that good and sufficient cause is presented to justify the Commissioner in extending this patent.

From the records and files of the office it will appear, that Samuel F. B. Morse, obtained letters patent of the United States, bearing date June 20th, 1840, and running for the term of fourteen years from their date.

These letters patent were surrendered to correct a defective specification, and new letters patent were issued, with a corrected specification, bearing date January 20th, 1846, and running for the term of 14 years, from the 20th of June, 1840.

These last mentioned renewed letters patent were surrendered on account of a defective specification, and new letters patent issued with a corrected specification, bearing date June 13, 1848, and running for the term of fourteen years from the 20th day of June, 1840.

Professor Morse duly filed his application for an extension of these letters patent so reissued, paid the requisite fee, and filed the required statement of receipts and expenditures.

Opposition was entered at the Patent Office, by Hon. R. H. Gillet, as Solicitor for Marshal Lefferts, by Freeman M. Edson, and by Robert W. Russell; and notice of such opposition communicated to this applicant on the 22nd of March last.

Further opposition was entered by David W. Baldwin, and notice thereof given to the applicant on the 5th of May last.

Testimony in support of the application of Samuel F. B. Morse has been taken at New York, Philadelphia, Baltimore and Washington, and duly filed in the office. Due notice of the time and place of taking of such testimony was served upon all the parties opposing, except, that the Philadelphia testimony having been taken before the notice of David Baldwin's opposition was received, he was promptly notified of the names and residences of the witnesses previously examined.

Professor Morse has also filed a copy of a decision of the United States Circuit Court for Pennsylvania, in the case of French vs. Rogers; and the correspondence of the office on the subject of the extension of his patent.

Rule 58 of the Patent Office requires that *"any person opposing the extension of a patent, must file his reasons in the Patent Office at least twenty days before the day of hearing as set forth in the notices published."* (See rules.) The same requisition is contained in the special order of the Commissioner as published in this case. (See published notice annexed hereto.)

Notwithstanding this rule of the office and the order of the Commissioner in this case, R. H. Gillet, Marshall Lefferts, Freeman M. Edson and Robert W. Russell, omitted to file their respective *"reasons."*

The applicant has, therefore, entered on the files, a formal protest against the right of these parties to submit any testimony or argument in opposition to this application.

Subsequent to the time of filing "reasons" R. H. Gillet filed the testimony of Henry O. Reilly taken on behalf of Marshall Lefferts at New York, the decision of the Circuit Court of Kentucky, in the case of Morse vs. O'Reilly, and the decision of the Supreme Court of the United States in the same case.

He also filed a notice of books and papers, intended to be read by him at the hearing of this cause. To this notice a formal objection and protest was entered on behalf of this applicant.

David Baldwin filed reasons setting forth substantially, that he opposed the extension of Professor Morse's patent, because, in 1845, he invented an improved plan of telegraph which he cannot use without infringing the patent of Morse. No evidence, however, has been taken or filed on behalf of David Baldwin.

The examiner in charge of this subject has reported in favor of the originality of Professor Morse, to the invention patented by him, and has directed the attention of the Commissioner to the propriety of a correction in the eighth claim of his patent by a reissue.

This being the state of the case, it will be contended on the part of the applicant, *that there is no party before the Office entitled further to oppose the extension of this patent.* Under, and subject to the Commissioner's decision on this point, *we desire to reply in advance to such matters as may possibly be presented in argument by opposing parties.*

First. There is no party entitled to be heard before the Commissioner, in opposition to this extension. Russell, Edson, and Lefferts, are not entitled to be heard because they neglected to file in the Patent Office their reasons, at least twenty days before the hearing. That they did so neglect to file their "reasons," is established by the records of the office.

By section 12, act of 1839, the Commissioner of Patents, is empowered to make all rules and regulations in respect to the taking of evidence to be used in contested cases before him as may be just and reasonable, and by act of 1848, section 1, the power of extending patents is vested solely in the Commissioner of Patents.

Under these acts, the Commissioner established the following general rule. "Any person opposing the extension of a patent must file his reasons in the Patent Office, at least twenty days before the day of hearing as set forth in the notices published." Moreover, upon the filing of the petition in this case, the Commissioner made and published an order, fixing the day of hearing, the day of closing and filing the testimony, and further ordering "that persons opposing this extension, are required to file in the Patent Office their *objections specially set forth in writing*, at least twenty days before the day of hearing." (See published notice.)

The Commissioner, having thus established this general rule and promulgated a special order, it is submitted, that all persons opposing this extension are bound to conform thereto. This rule is a wise and just one. It is intended to secure to the applicant seasonable information of the ground on which he is opposed, whether it be a matter of law or of fact so that the former may be met by argument or authority, and the latter rebutted by testimony.

It is a provision, moreover, essential for the protection of the applicant; otherwise, after his testimony had been closed, his argument filed, and his patent about to expire, he might be surprised by testimony or by arguments on points which he could easily and effectually have overcome.

A rule, then, of such importance, and so distinctly announced, should, we insist, be enforced; and we inquire *how* is it to be rendered available to the present applicant?

If the opposing parties had omitted to file their testimony before the day fixed for closing testimony, or to file their argument before the day of hearing, there could be no doubt that the Commissioner would reject both testimony and argument. Now, the same order which fixes a day for filing the testimony and arguments, requires, also, the "*reasons of opposition*" to be filed, and fixes a day for that purpose.

If, therefore, this rule of the office means anything, if the special order made and published by the Commissioner in this particular case is of any force, then, we respectfully contend, *that all testimony and arguments filed by opposing parties who have refused to comply with the provisions thereof, should be rejected.*

Had the opponents of an extension filed specific "reasons" of opposition, it is clear that they would have been restricted by the Commissioner to the reasons so filed; otherwise this rule would prove a snare to the applicant, leading him to suppose that all grounds had been specified, and inducing him to confine his proofs and arguments thereto, while the opposing parties might surprise and defeat the application by new and unexpected points taken at the last moment. For the same reason that a party filing his grounds of opposition would be limited in argument to the reasons so filed, should the party who neglects to file any reasons whatsoever, be precluded from submitting any argument in the case. By taking this course, the Commissioner will render this rule available to applicants. It will, moreover, be analagous to the practice of the Supreme Court of the United States and all other courts which require a brief of points and authorities to be filed at a fixed time. It will also be in analogy to the course pursued by the judge of the district court for the District of Columbia, when sitting as an appellate tribunal upon the decisions of the Commissioner of Patents, under section 11, act 1839—the judge expressly confining himself to the "reasons of appeal filed."

On the other hand, should the Commissioner in this case admit the testimony and argument of the opposing parties, notwithstanding their total neglect of his special order and rule, he would place them in a better condition than those persons who have complied with his rule; for the latter are restrained, in some manner, by the specific reasons filed, while the former, having no such restraint, could go into any extent of discussion they pleased.

This point is pressed in this case with some degree of earnestness, not from any disposition to insist on a technicality, but because it is thought that the disregard of this rule has not been altogether unintentional; and it is feared that such disregard might work INJUSTICE to the applicant.

Notice has been given by R. H. Gillet, that numerous books and papers will be read at the hearing of this cause; and the testimony of H. O'Reilly, (the only witness produced by Mr. Gillet, and within thirty hours of the time fixed on closing testimony,) has also been filed. The counsel for the applicant cannot know the purpose for which these papers or testimony are to be used, and must, therefore, prepare to meet every argument that can possibly be deduced from them. After exerting their utmost ingenuity, they may still fail to perceive and protect their client's case at the real point of attack.

R. W. Russell and F. M. Edson, have filed no papers as yet, and we are lost in still greater uncertainty as to the line of argument those gentlemen intend to adopt.

If the rules and orders of the Patent Office are to have any real force or be respected; if, indeed, a premium is not to be offered for their violation; if the rules of practice before other tribunals are to have weight here; if the course pursued by this office in other cases is to govern this case; then we submit, that the testimony and arguments of R. H. Gillet for M. Lefferts, the arguments of R. W. Russell and F. M. Edson should be rejected from this case by the Commissioner.

The opposition of David Baldwin stands on other grounds. He has, indeed, filed his "reasons" of opposition, asserting that he made a valuable improvement in electro-magnetic telegraphs in 1845, and that he cannot use it without interfering with Professor Morse's patent. But Mr. Baldwin has offered no testimony to establish the date of his invention, its nature, or its alleged superiority to Morse's invention. His case, therefore, rests solely on his own individual statement, and should not be seriously entertained by the Commissioner as a ground of opposition. Even if it be in all respects, as represented in his reasons of opposition filed, it would amount to no more than that Mr. Baldwin, being the owner of an improvement, could not use it without infringing. This objection might be urged against every extension that has ever been asked for. It is almost universally the case, that subsequent improvements are added to important inventions by others, besides the inventor. If Mr. Baldwin has really invented the valuable improvement he supposes that he has, let him patent it as an improvement, and Professor Morse, or his assignees, will cheerfully pay him for its use.

As already stated, he has omitted to prove that the public would be benefitted in any way by his improvement, or that he has been unable to introduce it by reason of Professor Morse's patent; and it is submitted that his statement, being unsupported by proof or explanation, *amounts to nothing* in the present case, and will be rejected by the Commissioner.

We have thus stated the grounds on which we suggest that the testimony and arguments of the opposing parties should be rejected. In order, however, to guard ourselves in case the decision of the Commissioner should be against us on these points, we proceed to conjecture, as far as we are able from the papers filed, the possible grounds on which they may file an argument against the application.

Vail's book, and certain papers have been filed by Mr. Gillet, containing a description of the inventions of Cooke & Wheatstone, Davy, Steinheil, and letters of Professor Morse. To these books a formal objection and protest has been entered on behalf of Morse: I. Because said Gillet has given no notice to the applicant for what purpose such testimony is offered; II. Because the copies of letters contained in the work of Alfred Vail are not competent proof of such letters; III. Because the loss or non-production of the original letters is not shown; IV. Because said books and papers, being in the nature of evidence, should have been filed or put in evidence as required by the rules of the office; and V. Because said books, papers, and patents are of date subsequent to the invention and application of Professor Morse for his letters patent, and are for different inventions from his.

In addition to the ground of objection before urged, the specific objections taken to these books and papers are urged upon the attention of the Commissioner, Mr. Gillet not having favored the applicant with any information as to the object for which they are to be read.

The question possibly intended to be raised by them is as to the originality of Professor Morse, as the inventor of the subjects claimed in this patent of 1840. The investigation of this question is one of immense scope. It has been four times judicially examined by the courts of the United States, and four times by the Patent Office, and in every instance decided in favor of Morse.

Morse's originality was investigated by the Commissioner of Patents, when he issued this patent in 1840, when he reissued the same patent in 1846, and when he reissued it again in 1848. It was reported upon by the Examiner in the present application. Judge Woodbury, at Boston, Judge Monroe, in Kentucky, Judges Grier and Kane, at Philadelphia, and finally the Supreme Court of the United States, have investigated it. Seldom has a question of fact been more thoroughly weighed.

The conclusion to which commissioners, examiners, and justices have arrived may be briefly summed up in the language of Chief Justice Taney, in expressing the unanimous opinion of the Supreme Court on this point. See decision of Supreme Court, filed by R. H. Gillett, page   :

"Waiving, for the present, any remarks upon the identity or similitude of these inventions, the Court is of opinion that the first branch of the objection cannot be maintained, and that Morse was the first and original inventor of the telegraph described in his specification, and preceded the three European inventions relied on by the defendants."

"The evidence is full and clear that when he was returning from a visit to Europe, in 1832, he was deeply engaged upon this subject during the voyage; and that the process and means were so far developed and arranged in his own mind, that he was confident of ultimate success. It is in proof that he pursued these investigations with unremitting ardor and industry, interrupted occasionally by pecuniary embarrassments; and we think that it is established by the testimony of Professor Gale and others, that early in the spring of 1837, Morse had invented his plan for combining two or more electric or galvanic circuits, with independent batteries, for the purpose of overcoming the diminished force of electro-magnetism in long circuits, although it was not disclosed to the witness until afterwards; and that there is reasonable ground for believing that he had so far completed his invention, that the whole process, combination, powers, and machinery, were arranged in his mind, and that the delay in bringing it out arose from his want of means; for it required the highest order of mechanical skill to execute and adjust the nice and delicate work necessary to put the telegraph into operation, and the slightest error or defect would have been fatal to its success. He had not the means at that time to procure the services of workmen of that character; and without their aid no model could be prepared which would do justice to his invention; and it moreover required a large sum of money to procure proper materials for the work. He, however, filed his caveat on the 6th of October, 1837, and on the 7th of April, 1838, applied for his patent, accompanying his application with a specification of his invention, and describing the process and means used to produce the effect.

*   *   *   *   *

"With this evidence before us, we think it is evident that the invention of Morse was prior to that of Steinheil, Wheatstone, or Davy. The discovery of Steinheil, taking the time which he himself gave to the French Academy of Science, cannot be understood as carrying it back beyond the months of May or June, 1837; and that of Wheatstone, as exhibited to Professors Henry and Bache, goes back only to April in that year. And there is nothing in the evidence to carry back the invention of Davy beyond the 4th of January, 1839, when his specification was filed, except a publication said to have been made in the *London Mechanics' Magazine*, January 20, 1838; and the invention of Morse is justly entitled to take date from early in the spring of 1837. And in the description of Davy's invention, as given in the publication of January 20, 1838, there is nothing specified, which Morse could have borrowed; and we have no evidence to show that his invention ever was or could be carried into successful operation.

"Now, we suppose no one will doubt that Morse believed himself to be the original inventor, when he applied for his patent in April, 1838. Steinheil's discovery does not appear to have been ever patented, nor to have been described in any printed publication until July of that year. And neither of the English inventions are shown by the testimony to have been patented until after Morse's application for a patent, nor to have been so described in any previous publication as to embrace any substantial part of his invention. And if his application for a patent was made under such circumstances, the patent is good, even if, in point of fact, he was not the first inventor.

"In this view of the subject, it is unnecessary to compare the telegraph of Morse with these European inventions, to ascertain whether they are substantially the same or not. If they were the same in every particular, it would not impair his rights. But it is impossible to examine them, and look at the process and the machinery and results of each, so far as the facts are before us, without perceiving at once the substantial and essential difference between them, and the decided superiority of the one invented by Professor Morse.

"Neither can the inquiries be made, nor the information or advice he received from men of science, in the course of his researches, impair his right to the character of an inventor. No invention can possibly be made, consisting of a combination of different elements of power, without a thorough knowledge of the properties of each of them, and the mode in which they operate on each other. And it can make no difference in this respect whether he derives his information from books, or from conversation with men skilled in the science. If it were otherwise, no patent in which a combination of different elements is used, could ever be obtained. For no man ever made such an invention without having first obtained this information, unless it was discovered by some fortunate accident. And it is evident that such an invention as the electro-magnetic telegraph could never have been brought into action without it. For a very high degree of scientific knowledge, and the nicest skill in the mechanic arts, are combined in it, and were both necessary to bring it into successful operation. And the fact that Morse sought and obtained the necessary information and counsel from the best sources, and acted upon it, neither impairs his rights as an inventor, nor detracts from his merits.

"Regarding Professor Morse as the first and original inventor of the telegraph, we come to the objections which have been made to the validity of his patent."—*Decision of Supreme Court.*

If it were necessary for Professor Morse to submit to the ordeal of another investigation on this question, we would not shrink from it, but would only ask that it should be thorough and deliberate. With the Examiner's report in favor of Professor Morse on this question, and with the combined testimony of justices and commissioners, it cannot be supposed that the Commissioner will now permit Mr. Gillet to open this question in an irregular and hasty manner.

Besides, the caveat of Professor Morse, filed in the archives in 1837, and his application for a patent filed, April 9, 1838, are anterior in date, and afford a satisfactory reply to the publications and papers referred. With these remarks we leave this point.

Mr. Gillett has filed in the Patent Office a copy of Judge Monroe's opinion in the case of Morse *et al. vs.* O'Reilly *et al.*, tried before him in Kentucky, and copy of a decree and opinion of the Supreme Court of the United States in the same case. It is conjectured by the counsel for the applicant that the intention may be to present to the Commissioner, two grounds of opposition based thereon, viz:

I. That Professor Morse's patent should have borne date from the date of a French patent, alleged to have been obtained in 1838, and that it would have expired before this, and cannot, therefore, be now renewed.

II. That the eighth claim of his patent has been held to be broader than the law warrants, and that the Commissioner ought not to extend a patent with such a claim.

As to the first point, viz: That Morse's patent should have borne even date with his French patent of 1838, and that it would have expired before the date of this application, and could not, therefore, be extended—it is submitted:

1. That there is nothing in evidence, or on record, to show the grant of a French patent to Morse, in 1838, or whether such patent was for the same or a different invention from that described in his patent of 1840.

Before the Commissioner could entertain this question, it would be necessary that the testimony on which such question would arise, should be fully presented before him. Now, neither the French patent, nor any copy thereof, nor of its specification, have been presented to the Commissioner. Hence it is contended that this point, if suggested by the opponents of the application should be rejected by the Commissioner as a point not legally before him.

2. For the purpose of the present application, the Commissioner is concluded by the date upon the face of the patent sought to be extended.

It has been often decided that a patent is *prima facie* evidence of the correctness of its contents. In this case there is no countervailing proof. By the face of this patent, it runs for fourteen years from the 20th of June, 1840. *That* is its *term; that* is the *term for which it was originally issued.* Now the act of July 4, 1836, section eighteen, contains but one restraining clause upon the power to extend patents, viz: *"Provided, however,* That no extension of a patent shall be granted after the expiration of the term for which *it was originally issued."*

The question, and the only question for the Commissioner, in this aspect of the case, is: For what term was this patent *originally issued?*

The original issue bears on its face the reply: *"for fourteen years from June 20, 1840."*

3. Even were the testimony sufficient to raise this question, and were the question to be now entertained by the Commissioner, then we submit that the Commissioner should decide such question in favor of the applicant; for the correctness of this term as expressed on the face of the patent has been three times sanctioned by the patent office, has been sanctioned by Justices Grier and Kane, by Judge Woodbury, by Justices Grier, Nelson, and Wayne, of the Supreme Court of the United States. That the term of the original patent was properly expressed on its face, was decided by Commissioner Ellsworth in issuing the patent in 1840; by Commissioner Burke, in reissuing the patent for the same term in 1846; and, again, by the same Commissioner, in reissuing the patent in 1848. In the case of French *vs.* Rogers, United States Circuit Court for the Eastern District of Pennsylvania, that court decided: "There is, therefore, no room for the questions, which were argued so elaborately of the proper interpretation of this proviso in the 6th section of act of 1839, and the 8th section, second clause, of act of 1836, which was also invoked, in any possible bearing upon the case of Mr. Morse. The proviso of 1837 must be interpreted by reference to the enacting words of the section which it limits; and the provisions of both the sections relate only to such patents as are applied for here after the issue of a foreign patent. But Mr. Morse's application here was before his patent abroad, (in no sense after it,) and his American patent was granted, therefore, under the general enactment of the act of 1836, not under any special proviso or exception whatever, and its term runs properly from its date."—Report of case of French *vs.* Rogers, pp. 3, 4. See report of case on file in the Patent Office.

Justices Grier, Nelson, and Wayne, of the Supreme Court, after a careful examination of the question, expressed the opinion "that the complainant's (Morse's) patent as renewed, contained a valid grant of the full term of fourteen years from its original date." See report of case of O'Reilly and Morse, in Supreme Court of the United States. And Chief Justice Taney, expressing the opinion of the majority of the court, said: "Nor is it void because it does not bear the same date with his French patent. It is not necessary to inquire whether the application of Professor Morse to the Patent Office, in 1838, before he went to France, does or does not exempt his patent from the operation of the act of Congress upon this subject. For if it should be decided that it does not exempt it, the only effect of that decision would be to limit the monopoly to fourteen years from the date of the foreign patent. And in either case the patent was in full force at the time the injunction was granted by the circuit court, and when the present appeal stood regularly for hearing in this court."

Against this mass of authority, there is but one opposing authority, that of Judge Monroe, in the case of Morse *vs.* O'Reilly, in the Kentucky circuit, a case filed, and probably to be referred to by R. H. Gillet in argument. It is to be especially observed, that while the Supreme Court affirm Judge Monroe in every other point, *four of those Justices* expressly decline to affirm him on this, and three of them protest against the position taken by him on this question.

4. Supposing, however, that the Commissioner should entertain this question, and, overlooking this mass of authority, were to consider that the patent ought to have been for fourteen years from the date of Morse's French patent; still it would be his duty to extend it for seven years, for then it would come within the words of Chief Justice Taney alluding to the possibility of such an alternative, and the Supreme Court would, when the question was presented to them, limit the monopoly to twenty-one years from the date of the French patent.

In this way the Commissioner would avoid assuming the decision of a question which four Judges of the Supreme Court declined to decide. The Commissioner's extension, if erroneously granted, could then be limited or nullified by the decision of that court. On the other hand, a decision on this point by the Commissioner against the extension would preclude the possibility of a revision or modification of his decision by any tribunal.

5. The Commissioner of Patents, Thomas Ewbank, when expressly applied to by Professor Morse, June 10, 1852, on this question, refused to entertain an application for an extension, saying, "*as this patent will not expire until 1854, an application now would probably be premature. It could not be considered so long in advance of the expiration of the patent.*" (See correspondence of Commissioner Ewbank and Professor Morse, filed in this case.) A solemn decision of a former Commissioner on this precise point, and upon which Morse has acted, would, (if not conclusive upon the present Commissioner,) at all events, be regarded as of the highest weight.

While the counsel for the applicant have entered into this

question of the French patent, they at the same time declare that they have done so only to anticipate a possible ground of opposition, and they protest against advantage being taken of any admission contained in this branch of their argument. They contend that the question has in no legitimate manner been brought before the Commissioner, and that if this point had been presented to their attention, either by notice or by the filing of reasons of opposition or by testimony, they would have submitted evidence to demonstrate that Morse's patent was correctly dated when issued.

The second point, for which it is presumed that the decision of the Supreme Court of the United States mentioned above, has been filed by Mr. Gillet, is, that the eighth claim of the patent of June 20, 1840, as reissued June, 1849, is broader than the law warrants, and that, therefore, the Commissioner should not extend a patent containing this claim. The attention of the Commissioner has been directed to the same point by the report of the Examiner on the present application, and, as these relate to the same matter, they may be discussed together.

The eighth claim of Morse's patent, as reissued, is in the following words:

"Eighth. I do not propose to limit myself to the specific machinery or parts of machinery described in the foregoing specification and claims; the essence of my invention being the use of the motive power of the electric or galvanic current, which I call Electro-Magnetism, however developed, for marking or printing intelligible characters, signs, or letters, at any distances, being a new application of that power of which I claim to be the first inventor or discoverer."

It was objected in the case of Morse vs. O'Reilly, that this claim was too broad. Four Justices of the Supreme Court, (Chief Justice Taney, Justices McLean, Catron, and Daniels) decided that it was not broader than Morse's invention; but that it was broader than by law he was entitled to claim. Three Justices decided that it was in no sense too broad. This claim had been previously sustained in the United States circuit court for Pennsylvania, and in the circuit court for Kentucky. Under these circumstances, the validity of this claim can hardly be said to have been decided by a majority of the Supreme Court, there having been two vacancies on the bench during the argument of this cause. The applicant, however, acquiesces in, and is desirous of conforming to, the decision of the majority of the Supreme Court as expressed by Chief Justice Taney; and with that view would ask the attention of the Commissioner to the following positions, as established by the Chief Justice's opinion:

First. That the eighth claim of Morse above cited is too broad, *inasmuch as it claims the use of electro-magnetism, however developed,* "for marking," &c.; but that Morse would be entitled to claim the use of electro-magnetism for recording *when developed by the mechanical combination,* which he has invented and described.

Second. That the patent of Professor Morse is not void by reason of the eighth claim being too broad, and that he has b
guilty of no laches in reference thereto.

Third. That Morse has a right to cure the said defect by a disclaimer.

First. The Chief Justice decided the claim of "Morse to be too
broad, inasmuch as it claims the use of electro-magnetism, *however developed*, for marking," &c., but that Morse would be entitled to claim the use of electro-magnetism for recording, when
developed by the mechanical combination, which he has invented
and described. This view of the Chief Justice will appear from
the following passages selected from his opinion :

"It is impossible to misunderstand the extent of this claim. He
claims the exclusive right to every improvement where the motive
power is the electric or galvanic current, and the result is the
marking or printing intelligible characters, signs, or letters at a
distance."

"Nor is this all. While he shuts the door against inventions of
other persons, the patentee would be able to avail himself of new
discoveries in the properties and powers of electro-magnetism
which scientific men might bring to light. For he says he does
not confine his claims to the machinery or parts of machinery
which he specifies, but claims for himself a monopoly in its use,
however developed, for the purpose of printing at a distance."

"No one, we suppose, will maintain that Fulton could have
taken out a patent for his invention of propelling vessels by
steam, describing the process and machinery he used and claimed
under it, the exclusive right to use the motive power of steam,
however developed, for the purpose of propelling vessels.

"Again, the use of steam as a motive power in printing presses
is comparatively a modern discovery. Was the first inventor of
a machine or process of this kind entitled to a patent, giving him
the exclusive right to use steam as a motive power, *however developed*, for the purpose of marking or printing intelligible character?"

*   *   *   *   *   *   *   *

"That is to say, he claims a patent for an effect produced by
the use of electro-magnetism distinct from the process or machinery necessary to produce it. The words of the act of Congress above quoted, show that no patent can lawfully issue upon
such a claim."

Such being the view of the court we proceed to the second
point.

Second. The patent of Professor Morse is not invalid by reason
of the 8th claim being too broad, and he has been guilty of no
laches in reference thereto.

All the justices of the Supreme Court agreed that the patent
of Morse was not invalid on account of the eighth claim.

Three of the justices were of opinion that no disclaimer need
be filed at all.

Justice Grier said:

"*Third.* Is it not true, as set forth in this eighth claim of the specification, that the patentee was the first inventor or discoverer of the use or application of electro-magnetism to print and record intelligible characters or letters? It is the very ground on which the Court agree in confirming his patent. Now, the patent law requires an inventor, as a condition precedent to obtaining a patent, to deliver a written description of his invention or discovery, and to particularly specify what he claims to be his own invention or discovery. If he has truly stated the principle, nature, and extent of his art or invention, how can the Court say *it is too broad,* and impugn the validity of his patent for doing what the law requires as a condition for obtaining it?"

The same judges were of opinion that no culpable delay had been committed by Morse in this matter.

Justice Grier said:

"Thus we see that it is only where, through inadvertence or mistake, the patentee has claimed something of which he was not the *first inventor,* that the Court are directed to refuse costs.

"The books of reports may be searched in vain for a case where a patent has been declared void, for being *too broad* in any other sense.

"Assuming it to be true, then, for the purpose of the argument, that the new application of the power of electro-magnetism to the art of telegraphing or printing characters at a distance, is not the subject of a patent, because it is patenting a principle; yet as it is also true that Morse was the first who made this application successfully, as set forth in this 8th claim, I am unable to comprehend how, in the words of the statute, we can adjudge "that he has failed to sustain his action on the ground that his specification or claim embraces more than that of which he was the first inventor." It is for this alone that the statute authorizes us to refuse costs.

"*Fourth.* Assuming this 8th claim claim to be too broad, it may well be said, that the patentee has not unreasonably delayed a disclaimer, when we consider that it is not till this moment *he had reason to believe it was too broad.*"

And C. Justice Taney said:

"It appears that no disclaimer has yet been entered at the Patent Office. But the delay in entering it is not unreasonable. For the objectionable claim was sanctioned by the head of the office; it has been held to be valid by a Circuit Court, and differences of opinion in relation to it are found to exist among the justices of this Court. Under such circumstances, the patentee had a right to insist upon it, and not disclaim it until the highest court to which it could be carried had pronounced its judgment. The omission to disclaim, therefore, does not render the patent altogether void, and he is entitled to proceed in this suit for an infringement of that part of his invention which is legally claimed and described."

Third. Morse has a right to cure the defect by a disclaimer if he choose.

Chief Justice Taney in the case before cited, said on this point: "It has been urged on the part of the complainants that there is no necessity for a disclaimer in a case of this kind. That it is required in those cases only in which the party commits an error in fact, in claiming something which was known before, and of which he was not the first discoverer; that in this case he was the first to discover that the motive power of electro-magnetism might be used to write at a distance; and that his error, if any, was a mistake in law in supposing his invention, as described in his specification, authorized this broad claim of exclusive privilege; and that the claim, therefore, may be regarded as a nullity, and allowed to stand in the patent without a disclaimer, and without affecting the validity of the patent.

"The law which requires and permits him to disclaim is not penal, *but remedial. It is intended for the protection of the patentee* as well as the public, and ought not, therefore, to receive a construction that would restrict its operation within narrower limits than its words fairly import. It provides, "that when any patentee shall have in his specification claimed to be the first and original inventor or discoverer of any material or substantial part of the thing patented, of which he was not the first and original inventor, and shall have no legal or just claim to the same,"—he must disclaim in order to protect so much of the claim as is legally patented."

"Whether, therefore, the patent is illegal in part, because he claims more than he has sufficiently described, or more than he invented, he must in either case disclaim, in order to save the portion to which he is entitled; *and he is allowed to do so when* the error was committed by mistake."

Morse having invented and described a specific combination of parts for the development of the motive power of electro-magnetism, to mark and record at a distance; and Chief Justice Taney, in expressing the opinion of the Court, having decided that the defect consists in "claiming *a patent for an effect produced by the use of electro-magnetism as distinct from the process or machinery necessary to produce it,*" and the Chief Justice having further decided that "he *may* and must *disclaim* in order to "protect *so much of the claim* as is legally patented," Morse has entered a disclaimer to so much of his 8th claim, as includes the use of the motive power of electro-magnetism for marking or printing intelligible characters, signs or letters at any distances, "*except when such motive power of electro magnetism is developed by the combined action of a long galvanic circuit, a contrivance for closing and breaking the circuit, an electro magnet, armature and spring, or their equivalents, substantially as set forth in said specification.*" (See disclaimer of Morse filed.)

This disclaimer has been duly entered by Professor Morse, in accordance with the statute in such cases made and provided. In order to meet the precise view of the majority of the Supreme Court, as expressed by Chief Justice Taney, much care has been observed in its preparation. The peculiar form of the disclaimer and petition adopted is similar to one entered by Elisha Foote, on his patent for a stove-regulator. That disclaimer underwent a full examination in the United States Circuit Court for the Northern district of New York, and a subsequent revision in the Supreme Court of the United States; it was sustained by the latter tribunal, Judge Curtis delivering the opinion of the court, and but one judge dissenting. See the case of Foote vs. Silsby, as reported in 1 Blatchford, 447, and the same case on appeal in 14 Howard, page 219.

The entry of this disclaimer by Morse, it is submitted, entirely obviates any objection arising from the decision of Chief Justice Taney, upon the 8th claim, and complies with the suggestion of the examiner as to the necessity of a surrender and reissue.

We have thus shown that neither of the parties who have entered opposition in this case, are entitled to be heard on the present application, and that the testimony, books, and documents referred to by Mr. Gillet, should be rejected from the case. Subject to the Commissioners decision on these questions, and with a view to fortify ourselves against any arguments of said opposing parties, that the Commissioner might determine to receive, we have shown that the French patent question is not properly before the Commissioner; that on the case as presented, the Commissioner has full authority to extend said letters patent, and that the weight of previous authority enforces this view. We have also set forth the objection of the Chief Justice of the Supreme Court, to the 8th claim—shown that Morse has a right to disclaim, and referred to the disclaimer filed by him as overcoming the objection made by that court. Before passing from this branch of the subject, we desire to repeat our protest against being prejudiced by any positions or arguments which we have thus advanced. Having been forced to pursue this course by the neglect of our opponents to comply with the rules of the office, we submit that our arguments and positions under such circumstances should be regarded by the Commissioner at the hearing, solely as matter argumentative and by way of rebuttal, and not as conclusive upon us as to any point not otherwise before him.

We pass to the case as presented on behalf of the applicant. The question of novelty and patentability having been established by the report of the examiner, the repeated decisions of commissioners and courts, and finally by the unanimous opinion of the highest court of appeal in the country; it only remains for us to analyse the statement and testimony filed on behalf of the applicant, and to establish from these that the patentee without neglect on his part, has failed to obtain from the use and sale of his invention, a reasonable remuneration for the time, ingenuity, and expense bestowed upon it, and the introduction thereof into use.

2

1. *What, then, is the value of this invention?*

2. *What degree of diligence has he displayed,* in perfecting his invention and introducing it into use?

3. How has the patentee been remunerated?

The value of the Magnetic Recording Telegraph cannot be estimated in dollars and cents. Its vast importance and utility present themselves intuitively to every mind conversant with the daily business of men and the ordinary casualties and affairs of life.

Men of the most extensive minds and enlarged views have been consulted in order to fix its pecuniary value, and they unite in declaring the impossibility of making such an estimate. All that could be done in this regard was to produce as witnesses, the leading merchants, manufacturers, and brokers, public magistrates, government officers in charge of public departments, the officers of railroads, and others resident in New York, Philadelphia, Baltimore, and Washington, and constantly using Morse's magnetic recording telegraph. Over twenty such witnesses have been examined, and the question has been propounded to each of them—"what, in your opinion, is the value of the Morse magnetic recording telegraph to your own particular business or department? and what is its value to the mercantile, manufacturing, and social interests of the community at large?" In answering this question some have stated their opinion in general but, emphatic terms; others have mentioned specific cases within their knowledge, illustrating its actual practical value; some have stated amounts which they would be willing to pay, rather than be deprived of it; others have dwelt upon cases where no moneyed value could possibly be thought of—cases of sickness and death, of an otherwise ruined credit saved, the speed of mails increased, accidents and death prevented, police regulations rendered efficient, destruction of property by fire and by sea prevented, and the progress of civilization and enlightenment advanced. It would be impossible, without reading the whole testimony to exhibit this subject with clearness and force to the mind of the Commissioner, and only a leading outline of the facts presented will be here attempted.

George Manly, a broker of Philadelphia, says; "We make daily use of this, sometimes as often as fifteen or twenty times a day; the object being to facilitate our buying or selling stocks in this or the New York market, or both. I do not think we could get along without it at the present time; it seems to me impossible that we should. * * * * * In times previous to the operation of the telegraph, the correspondence that is now had by telegraph was had by letter, and the business has increased 2 or 300 per cent. in consequence of the telegraph."

The testimony of this witness shows the fact that by means of the telegraph, business men, with a given capital, are enabled to increase their business by an immense per centage.

F. H. Carter, of the firm of Josiah Lee & Co., of Baltimore, confirms this view, and fixes ten per cent. of the commercial capital of the country as a small estimate of its value.

G. W. Clark, the head of one of the most extensive banking exchange houses in the country, having branches in Philadelphia, New York, and St. Louis, makes daily use of the telegraph, and says: "It is impossible to fix its value in dollars and cents; but it is of *immense value*." He mentions an instance by way of illustration, in which one thousand dollars was made by him in a single transaction by means of this telegraph, which could not have been made otherwise. He oftentimes gets ten or a dozen telegraphic messages a day, at Philadelphia, and the aggregate number at the St. Louis and New York offices would be more than double that number. He also alludes to its value in facilitating the operations of exchange between various places, and the saving of time and interest thereon. (See Philadelphia testimony.)

When the immense amount of exchanges daily going on between the principal cities of this country is considered, and when it is remembered that the regulation of this exchange was formerly effected solely by mail communication, and is now almost entirely effected by telegraph, it will be readily perceived that the aggregate of interest saved throughout the country during the time formerly occupied in communication, must be immense.

James H. Carter, a banker of Baltimore, (one of the firm of Josiah Lee & Co.,) also explains the value of the telegraph in this connection. He says:

"In the banking business of the firm of which I am a partner, the uses of the telegraph are varied and most important. In the ordinary course of the mail, a draft maturing in Cincinnati this day, Thursday, could not be availed of here, if paid there, until next Monday, when if no protest come to hand it could be drawn against. The telegraph makes it available here as soon as it is paid in Cincinnati; and there is, in this particular case, a saving of four days interest, besides the immediate availability of the capital otherwise. Three-fourths of the entire stock business between Baltimore and New York is transacted through the medium of the telegraph. It defeats fraud and forgeries by communicating the particulars thereof throughout the country, enables us to stop the payment of suspected drafts in distant cities, to ascertain the standing of parties in remote places upon the instant, and generally in all the relations of our business increases its efficiency and power."

Joseph Lea, of the firm of Hacker, Lea, & Co., an extensive commission house of Philadelphia, in addition to the advantages of the telegraph mentioned by others, alludes to its value in placing the manufacturer in immediate communication with the commission merchant. Thus, it is well known, that our great manufactories and print works are located at Fall River, Lowell, and other places distant from New York, Boston, and Philadelphia, where the merchants reside who sell the manufactured goods.

By means of the telegraph, the merchant is enabled to control and regulate the manufactory with the same promptness and efficiency that he could do were it located in immediate contiguity to his counting house. On this point, Mr. Lea says:

"We are in the habit of using it (the Morse telegraph) almost daily in expediting the production of various styles of goods, and also in correcting errors in their assortment by which their market value is considerably enhanced. We are also in the habit of receiving orders by telegraph from remote points by which several days are gained in the delivery of the goods. The telegraph is daily used for this purpose by almost every merchant in the country, and with a manifest pecuniary advantage to their business."—(Philadelphia testimony.)

One other service often rendered by this telegraph is the protection of the credit of individuals, a matter to mercantile men, second only in importance to the preservation of life and health. On this point Mr. Lea says:

"It is invaluable to the mercantile community in correcting errors in the remission or delay of money by mail. I can give an instance of several thousand dollars having been missent by a large house in Cincinnati to New York instead of Philadelphia, where their note falling due would have been protested, and in the critical position of the market would have subjected them to discredit. In thus enabling houses to avoid discredit, which frequently arose before the introduction of the telegraph, from the delays of mails and other causes, we consider that private credit has has been placed upon a more secure basis, greater activity imparted to business transactions, and the risks of business essentially lessened. In these respects no estimate can be formed of the value of the telegraph to the mercantile interests of the country."

John E. Addicks, an extensive merchant of Philadelphia, testifies to the same point, and expresses the opinion that the mercantile interests of Philadelphia had better pay one or two hundred thousand dollars a year than be deprived of the use of this telegraph; and this, he thinks, would be an estimate far within the mark.

Robert Hoe, of the firm of R. Hoe & Co., of New York, extensive manufacturers of printing presses and saws, confirms the statements of the Philadelphia witnesses as to the enormous and incalculable value of this telegraph to the commercial and social interests of this community. This witness also mentions the ease of the employment of a private Morse telegraph line by his firm, about two miles in length, and connecting their workshop and factories with their counting-house. After a constant use of this private line for three or four years, he states, as the result of his experience, that "it enables them to conduct a large establishment, remote from the centre of business, with a despatch and convenience which would otherwise be impossible." (See New York testimony.)

Cyrus W. Field, a merchant of New York, in frequent use of this telegraph, corroborates the foregoing, and says: "It is like the application of steam power—the commercial community cannot dispense with it. A message of a few words is often of immense value." (See New York testimony.)

Of the same tenor is the testimony of Peter Cooper, of New York, George R. Dodge, of Baltimore, George W. Dobbyn, president of the Susquehanna and Tide Water Canal Company, William Boze, editor and publisher of "The Baltimore Daily Advertiser," Louis McLane, of Baltimore, B. B. French and George Vail, of Washington.

The value of this telegraph in the ordinary social affairs of life is of the highest importance, and demands a moment's attention.

Joseph Bench, an operator upon the New York and Washington line of telegraph, arranged under Morse's patent, says that a record kept for a week shows that the average number of messages in matters of sickness and death sent from the New York office south is over six per day, and that a considerable number are also received. This is but from a single office and in a single direction.

The extent to which this telegraph is used and the number of persons who avail themselves of it, is shown by the testimony of Joseph Bench and Joseph Sailer. Bench tells us that during last month 5,460 messages were sent from, and 5,468 were received at, New York over the New York and Washington line alone. (See New York testimony.)

Joseph Sailer states that the aggregate number sent and received at Philadelphia, over the same line during the last month, was 11,123.

Joseph Sailer states that the aggregate receipts by the Morse Magnetic Telegraph Company, for messages between New York and Washington and intermediate places during the past year, is $113,080 33, showing what amount the public are willing to pay for the use of this instrument on one line of about 240 miles length.

Perhaps one of the most useful applications of the telegraph is to the police service of the country, for the arrest of criminals and the recovery of stolen goods and money. Mr. Bulkley, lieutenant of police at Philadelphia, exemplifies its public value as an arm of justice and a safeguard against crime and its consequences. He mentions an instance of a robbery of $3,000, and the prompt arrest of the criminal and recovery of the money by the aid of the telegraph. Another similar case where the amount was $2,300; a case of the recovery of $1,000 or $5,000; one of the robbery of a bank of $65,000, and other instances, in all of which the parties would have escaped and the money been lost but for the telegraph. He mentions these only as illustrative of its value to the police operations and interests in Philadelphia, and says that his use of the telegraph is frequent, and he thinks it to be invaluable to police operations and interests. (See Philadelphia testimony.)

This, it is to be observed, is the result achieved by it in particular cases in a single city, but this becomes startling when considered in reference to all the cities of our country.

Look, also, at its value in connexion with municipal regulations for the extinguishment of fires. T. P. Shaffner thus speaks of this subject:

"I have examined the fire-alarm telegraph of the city of Boston, composed of the Morse system of telegraphing. The superintendent of said telegraph, (which is under the control of the Boston city government,) informed me that said telegraph had saved at least some millions of dollars per annum—that no fire could possibly get under much head way before the engines would be on the spot.

"Any person (man, woman, or child) can at any time, day or night, within one minute, notify a man at a central depot where the fire is, who then rings several large bells stationed at different parts of the city, informing the firemen where the fire is.

"If the man at the central depot is not in the room when the fire alarm is given, he is notified by a bell; and when he returns to the machinery he is enabled to see where the fire is, the fact being written on the paper by the self-regulating machine.

"This can only be accomplished by the Morse system." (New York testimony.)

* * * * * *

The General Government derives great advantages from this telegraph, not only in the ordinary conduct of its business, where its officers (like other individuals) are compelled to use it, but in certain departments, experience has demonstrated its pre-eminent value. For example, in the Post Office Department for the regulation of the mails, and in the War Department in the movement of troops and of supplies. On the former subject Mr. Murphy, chief clerk of the Philadelphia post office, testifies:

"The Philadelphia post office does make use of it very frequently. In case of important mails failing to reach here we telegraph to the important point, such as New York or Boston, to ascertain the reason, and to know if it was sent. We had a case in point about two weeks ago, in which the mail from London for Philadelphia failed to reach here; it created a great deal of trouble among the financiers and others in not receiving their letters, and, in order to satisfy them as to the cause, I telegraphed to Boston, and, from the answer, was enabled to satisfy the parties that the mail was left in London; this was all done in the space of an hour at most. In case of an accident to the mail leaving New York, despatches are sent by telegraph not to wait for our own mail, but to send the southern mail off without delay and at the regular time, sometimes enabling us to save a southern connexion and a day in time between here and New Orleans, and preventing connections failing at points south of us. I have been connected with the Post Office Department for twelve years. * * I consider it of the greatest utility, and the postal arrangements would be very much retarded if it were not in existence. In case of a demand for mail-locks or mail-pouches at any one point, its use would save a day in their delivery. It is oftentimes used to prevent the negotiation of stolen notes and other mail property, and you may employ the strongest term you like as expressive of its value."

* * * * * * *

Its importance to the War Department, in the movement of troops or supplies, is mentioned by Mr. Shaffner, and is, indeed, almost self evident.

In case of a war with Spain, or an attack on Cuba, who can estimate the importance of an instantaneous communication between the Department, at Washington, and the military or naval depot, at New Orleans.

Another obvious and extensive use of the telegraph is in the coast lines of telegraph. Lines of telegraph extend from New York to Sandy Hook, on the coast, and from Philadelphia to the Breakwater. In case of Marine disasters, the underwriters at New York or Philadelphia are instantly apprised thereof, and steamtugs, pumps, lighters, &c., may be instantly dispatched to the relief of the vessel in peril.

The amount of property saved in this way, during the last winter must have been enormous.

The utility of the telegraph to railroad companies is, perhaps, more direct and apparent than the other uses alluded to.

John Tucker, the President of the Philadelphia and Reading Railroad Company, says, his company employ this telegraph constantly, and sets forth and illustrates its value.

"The chief value to us is obtaining early information in regard to accidents, but we also use it very extensively to obtain or impart information as to all business of the company; for instance, our superintendent at our coal depot at Richmond, reports every evening, to the car distributor in the coal region, the exact number of cars detained loaded by each operator, which is the basis for the distribution of cars the next morning, as the parties who detain loaded cars are not entitled to empty ones; this is very important information for this car distributor to have; the same information from this and other points is communicated to me daily from that and other points, and it gives me perfect information about the entire business of the company, which business, per mile of road, is the largest of any railroad in the world.

"In case of accident or detention of the trains, information can be instantly communicated to the proper station, by which means the cause of the detention is immediately removed, and thus not unfrequently a loss of business equal to from four to seven thousand dollars is avoided.

"A strike occurred yesterday, among the brakesmen and conductors along the line of the road, which they communicated to me by said telegraph, stating their reasons. I replied by the telegraph, requesting them to resume their duties, and give me time to investigate the causes of their complaints, and they at once responded 'that they would do so.' If we had lost even the one day's business for want of the proper communication, the receipts of the company would have been diminished ten thousand dollars, and the nett profits more than five thousand dollars; the length of our road is ninety-three miles."—*Philadelphia Testimony.*

Samuel M. Felton, President of the Philadelphia and Baltimore Railroad Company, mentions the following illustration of its great value:

"On one particular time the Susquehanna river was obstructed by ice, and we had a railroad track across the river on the ice, which we expected every moment would break up. At that time I was in the habit of receiving and sending messages by telegraph to Havre de Grace, as often as once an hour, or oftener, and by which I was made acquainted with the precise condition of the ice in the river, and prepared to give directions to each train as it left the depôt here or in Baltimore. When the ice finally broke up, a dispatch was sent to Baltimore, to the agent, to send the train by steamboat line by way of Frenchtown and New Castle, as it was no longer possible to cross the ice at Havre de Grace, or by ferry over the Susquehanna. Without the telegraph, this information could have been obtained only by running locomotives to that river from Baltimore and this city, and this would have been attended with delay, comparatively great delay and expense. By means of the telegraph, I have often conversed with our superintendent at Baltimore or Havre de Grace, by the hour at a time, and made arrangements with as much facility as though we were both present in the same room. I have, also, arranged a schedule, for the running of trains, with the superintendent of the Baltimore and Ohio railroad, he being at the time at the telegraph office, in Baltimore, and I at the telegraph office here. We, also, have the earliest possible intelligence of accidents and detentions along the line of the railroad, and are able to give directions to the several trains and station agents along the road, how to proceed under the circumstances. In this way much time and money are saved to the company, and the traveling public subjected to the least possible delay.

This witness adds, that it would be as difficult to estimate its value in reference to railroads, as to estimate the value of a good education to an individual.

From the testimony of George R. Dodge, it appears that the Baltimore and Ohio Railroad Company, paid $28,000, in money, towards the construction of a line of telegraph, along their road, and incurred an annual expense of $16,000, in keeping up stations, &c., merely for the privilege of telegraphing in reference to the business of their road, when it was not otherwise occupied. The last report of the New York and Erie Road, shows that a line of telegraph has been built along their road, costing $50,000, and involving an annual expense of $20,000, and, yet, the report speaks of this—"As has been previously stated, this company has in operation four hundred and ninety-seven miles of telegraph, used exclusively for its own business, and fifty-two offices and sixty-five operators employed. No expenditure which has been made in this work has proved more profitable than that made for this purpose. It has added to the safety of passenger*, and has given a feeling of security to the managers and operatives of the road, against a large class of accidents, to which, without it, they are peculiarly exposed."

It is impossible fully to develope this branch of the subject. We cannot turn in any direction without discovering some useful and valuable application of this invention. A thought just occurs to us as to the great value of it in connection with the const survey, in determining the difference of longitudes, and we doubt not that we have overlooked many of its most important applications.

T. P. Shaffner thus sums up his experience of its value.

"During my intimate connection with the Morse telegraph for the past seven years, I have seen the great benefits of his system realized by the community. I have seen life saved, criminals arrested who would have escaped, fortunes accumulated and property saved from destruction by the elements of fire and water."

As to the ingenuity displayed by him, and the relative merit of his invention compared with others, it is only deemed necessary to cite the deliberate opinion expressed by Mr. Justice Grier on this point. "The word telegraph is derived from the Greek, and signifies to "write a far off, or at a distance." It has heretofore been applied to various contrivances or devices, to communicate intelligence by means of signals or semiphores which speak to the eye for a moment, but in its primary and literal signification of *writing, printing* or *recording* at a distance, it never was invented, perfected or put into practical operation till it was done by Morse. He (Morse) preceded Steinheil, Cook, Wheatstone and Davy, in the successful application of the mysterious power or element of electro-magnetism, to this purpose; and his invention has entirely superseded their inefficient contrivances. It is not only "a new and useful art," if that term means anything, but a most wonderful and astonishing invention, requiring ten fold more ingenuity and patient experiment, to perfect it than the art of printing with types and press as originally invented."

We have thus given an outline of the value of the Magnetic Recording Telegraph as proved in this case. Morse having been declared by the Supreme Court of the United States to be the "*inventor of the telegraph*," to be the inventor not of the recording telegraph, but to be in advance of Cooke & Wheatstone, Davy & Steinheil, as the inventor of any form of electro-magnetic telegraph, he is entitled to a remuneration in some manner proportionate to the value of the invention thus proved. Neither is it to be considered that If Morse had not invented the telegraph, some other person would. Grant that another would have invented the telegraph, then that person would have been entitled to this remuneration instead of Morse. The Government grants a monopoly for fourteen years as a reward to the first man who invents a machine or process, &c. This monopoly is the reward held out in the race of invention.

If ten men invent the same thing within the same year, he who first invents obtains the exclusive right for fourteen years, while the other nine obtain nothing, although they may be equally original, ingenious, and industrious with the one who obtains the patent. It is the policy of Government thus to look upon inventors as contending in a race and grant the reward to the winner. All who run a race exert nearly the same energy, but he only who wins receives any reward for this toil. This policy of Government secures perseverance and despatch. What men find to do they must do quickly. It may be that the same invention would have been made by another within a week, but it might happen that it would not have been invented for ten, twenty, or thirty years, and it is upon the latter hypothesis that Governments proceed in granting patents. And the extension of a patent rests upon the same ground, not being a new grant, but an equitable enforcement of the old.

A patent when obtained is a consideration paid by Government to the inventor, 1st, for having given the public an invention; and, 2d, for having given it to them before any one else. If it turns out that the public will not use an invention for several years, or if they steal and use it without paying for it, or if they take advantage of the inventor's poverty and extort it from him, then the Government deems the consideration which they have given to have failed, and, if he be not in fault, make that consideration what they originally intended it to have been, by extending that grant for a longer period.

We proceed to the next point.

As to the inventor's diligence in perfecting and introducing this invention into use, this case presents itself to the Commissioner in the strongest possible light. The history of this invention is one of fourteen years of toil and devotion by Morse to his project amid difficulties, privations, sneers, and disappointments; to this succeeded a period of eight years of treachery on the part of agents and associates, public defamation, infringements and lawsuits. This has occupied the time almost down to the very hour of filing this application. Finally, when worn down with these struggles of the fourteen years of the life of the patent and the sixty-third of the inventor, the highest judicial tribunal in the world unanimously awarded to him the proud honor of being the original "inventor of the electro-magnetic telegraph." The detail of Morse's diligent labors and toils to perfect and introduce his telegraph are clearly and truthfully set out in his statement on file. The prejudice and opposition of the public, the indifference of Government, and the aversion manifested by capitalists to his scheme, are set forth in the testimony of his brother, Sidney Morse, Hon. Louis McLane, B. B. French, George Vail, and George R. Dodge, on file in the case.

The first line that Professor Morse could obtain sufficient pecuniary aid to construct was the Washington and Baltimore line, finished in the spring of 1844, about 40 miles long. In the year 1846, the line between Washington and New York, of about 240 miles, was completed. What was Professor Morse's diligence up to that time? Sidney E. Morse testifies as to the pecuniary circumstances of Professor Morse from 1832 down to the time of the introduction of the invention and of his diligence in the matter.

"At the close of the year 1832, when he returned from Europe, he was without pecuniary means. My younger brother and myself provided him with his studio for painting in the city of New York, and for several years I supposed and believed that his principal means of subsistence were furnished by us, and he continued to be almost without other income until he received the $2,000 allowed him by the Government of the United States for superintending the erection of the telegraph between Baltimore and Washington, in 1844.

My own business was such as to engross my time and attention almost entirely, but I know that he was engaged in experiments, diligently and earnestly, when not prevented by the necessity of providing for his necessary wants, from the year 1832 down to the time of the perfection and general introduction of his telegraph."

The labors of Professor Morse before Congress, to obtain aid from 1838 to 1843, are narrated by B. B. French, a witness in the case:

"Professor S. F. B. Morse came to Washington in the winter of 1837-8, with his invention of a magnetic telegraph, and exhibited it to such members of Congress and other persons as chose to examine it, in the room of the committee on commerce of the House of Representatives of the United States. It was then a cumbrous affair, as compared to its present appearance. The Professor had ten miles of wire in a coil upon a cylinder, through which he passed his galvanic current. I was introduced to Professor Morse, at Washington, and took a deep interest in his telegraph, firmly believing from his explanations, and my own examination and study, that it would work. The committee on commerce made a favorable report on the subject, but nothing further was done at that session. The Professor was very anxious to obtain an appropriation to test the merits of his telegraph, and urged it upon Congress with great zeal.

"In May, 1838, Professor Morse sailed for Europe, his object being, as he informed me, to secure patents from the Governments of that portion of the world for his invention. He returned to America prior to 1842, and during the winter of that year, came again to Washington, with his invention greatly improved, very nearly in the shape that it now is. He placed a register in the committee of commerce of the House of Representatives, and another in the committee room of naval affairs of the Senate, they being the south and north rooms of the center building of the capitol, and messages were sent from one room to the other over wires connecting the two instruments. At that time I was daily with Professor Morse, and he and his friends were as diligent as men could well be, not only in endeavors to obtain from Congress an appropriation to test the invention, but in endeavoring to enlist private enterprise in it. In March, 1843, a bill passed appropriating thirty thousand dollars to test the invention."—*Washington testimony.*

The same point is established by the evidence of George Vail, who also shows the existence of a strong prejudice in the minds of the community against the practicability or utility of the invention. He says:

"The great mass of intelligent men regarded the magnetic telegraph as a mere visionary scheme, from which nothing useful could ever be realized; and Professor Morse was looked upon as a clever and ingenious man, but scarcely of sane mind, because he proposed the telegraph as a medium for the regular communication of intelligence between distant places. In view of the peculiar difficulties encountered in bringing this invention into public and general use, mainly, if not wholly attributable to the absence of public confidence in its practicability."

The difficulties in overcoming prejudices and organizing a line from Baltimore to New York, are further testified to by John H. B. Latrobe and Lewis McLane, late Minister to England. The witness last named knew Professor Morse in Europe, and thus narrates the perseverance of Morse, as observed by him, and the difficulties which he there had to encounter:"

"I first knew anything about it from Mr. Morse himself, at the time he was engaged in maturing his discovery, and afterwards when he was endeavoring to procure an English patent. He encountered much mortifying opposition. In England he was looked upon by many as an adventurer, and it fell in my way while there to speak of him as he deserved, as a gentleman, an artist, and a man of genius, to persons who I was desirous should have a true estimate of him. A previous acquaintance was thus continued, and I became aware of Mr. Morse's object, and the persevering assiduity which he devoted to it.   *   *   *

"From all my intercourse with, and observation of Mr. Morse, until the line was in successful operation between Baltimore and Washington, he appeared to be perseveringly and zealously engaged in urging his invention into use. He seemed to have no other object, and everything I saw in connection with him, led me to consider him as exclusively devoted to the task of inducing the public to take his invention into favor. I saw no want of diligence on his part in connection with his introduction into general use."

Not until 1846 did Morse succeed in overcoming public incredulity and satisfying the mind of the community that his telegraph was a work of practical utility. Having thus shown his diligence from 1832 to 1840, when he obtained his patent, from 1840 to 1846, when he introduced it, we need no testimony to show his continued perseverance from that day to this. It is a matter of public notoriety. Starting from the short wire between Baltimore and New York, this telegraph has spread North, South, and West with a rapidity unparalleled in the history of invention. Within four years the line of telegraph wire in this country had increased from 240 miles to 23,280 miles, and had stretched itself around the Atlantic coast and the Gulf from Halifax to New Orleans, and the Balize, passed across the country from Boston through St. Louis, and thence down the Mississippi again to New Orleans. On the Northwest it had crossed the Alleghanies and reached to St. Joseph's, in Missouri, and to the extreme verge of civilization, intersecting and penetrating the country in every direction. For an accurate list showing the extent to which lines of telegraph had reached in 1850, the Commissioner is referred to the official abstract of the Seventh Census, as reported in 1851, pages 112, 113. This length of lines of telegraph has since been increased to about 30,000 miles, and preparations are making by Professor Morse and American capitalists to sink the great Atlantic sub-marine cables. Europeans, in the meanwhile, looked on with wonder and admiration at the spread of this invention in America. The telegraph in England had increased to less than 4,000 miles, and in France and on the Continent its growth had been slower. (See Official Census on same page.)

An English writer in Newton's Journal for 1850, thus comments on the progress of American telegraphs up to that time. Newton's Journal, vol 37, p. 205:

"The Ohio, Indiana, and Illinois Telegraph Company, we are told, has already 1,000 miles of line in working order, and this through a country covered, not with human dwellings, but dense and almost impervious forests; thus exhibiting another and a striking instance of the mode in which the latest discoveries of modern science are made subservient to the wants of communities at the very commencement of their existence. Half a century since, wild beasts, and still wilder Indians, wandered over the lands now traversed in perfect security by these frail wires, the mysterious agents by which human thoughts and desires are made to travel, in fact, as rapidly as they are conceived. This transition from a wild and barbarous condition to that of the most elaborate civilization, has not been gradual, but instantaneous. Civilization has not here at first dimly dawned, then slowly, gradually working its way against opposing difficulties to its ultimate perfection, but has at one bound leaped into life, surrounded with every appliance and means which the existing knowledge of man has devised for ministering to his wants and his enjoyment.

*   *   *   *   *   *   *

Let any one place before himself the map of the United States, and trace the distance that intervenes, for example, between the cities of New York and Cincinnati, and then let him regard the character of the country all along that extended line—the boundless forests, the wide, deep, and numerous rivers, the lofty mountains that must be traversed by the wires which connect the two cities we have named. Again: from New Orleans to New York the route is just as difficult; yet here again we see that the obstacles have not prevented communication, and, to our astonishment, we discover that there is a more rapid and continued correspondence between the people residing in Louisiana, New York, and Massachusetts than between London and Bristol. Is there not here cause for wonder and regret?

"When we remember that the discovery thus employed was made in England!* that the people of England have quite as great a need of rapid intelligence as those who dwell in America —that we have greater wealth, an equal intelligence and energy, we are driven to ask, Why have we not made the same use of this great invention?"

The history of the progress of the American telegraph is the best evidence of the diligence of the inventor in introducing his invention into use.

Before passing from this branch of the case, we desire to allude to the testimony of H. O'Reilly, taken at New York by R. H. Gillet, on behalf of Marshall Lefferts.

The right of M. Lefferts to submit any testimony in the case is denied by us, as before stated; but, for greater security, that testimony should, perhaps, receive a brief explanation.

Mr. O'Reilly alleges, in his evidence, that he made a contract with Morse and others in reference to the construction of certain telegraph lines; that this contract was violated by Morse and others; that they brought suit against him (O'Reilly) as alleged, for violation of this contract, and failed to sustain such suits; that, in the opinion of said O'Reilly, if Morse had permitted him to carry out his contract, Morse and his associates would have realized $500,000, or three-quarters of a million. This mere opinion the said O'Reilly bases upon some statement or paper or affidavit, in which he believes Mr. Kendall alleged that the profits which Mr. Morse and his associates had made at or prior to that time—the summer of 1848 or 1849—amounts to about $200,000, and upon other statements which he heard Mr. Kendall or some other persons make at sometime and somewhere.

Now, Mr. O'Reilly admits, on cross examination, that he can only swear to Morse prosecuting one suit against him, and that such suit went no further than an application for an interlocutory injunction; he cannot swear that Morse ever derived one cent from any contract made with him. He admits that he (the said O'Reilly) did controvert and deny the validity of the Morse patents. O'Reilly admits that, although he was himself entitled to one-fourth part of the stock in the lines built by him, yet, that he

* This note of interrogation was introduced by the editor of Newton's Journal.

has realized no profit himself, and that his whole property is in the hands of receivers. He admits that suit was commenced against him for infringing the Morse patent, in the United States circuit court for Kentucky, and a perpetual injunction awarded against him, commanding and enjoining him against infringing on the Morse patent; that this injunction was affirmed by the Supreme Court of the United States, is undissolved and in force against him. He further admits that, believing himself to have the control of certain shares of stock belonging to Professor Morse, he executed a letter of attorney to F. O. J. Smith to collect these shares at a time when he knew Smith to be hostile to Morse, and that he did this without Morse's consent. This witness thus shows himself to be deeply interested against this extension of the patent by reason of the subsisting injunction. He shows himself to possess a strong animosity to Professor Morse; that he has not been able to take care of his own property; and his vague and hearsay account of receipts, such as $200,000 made by Mr. Kendall in 1848, are mere matters of opinion, and are so extravagant as to need no remark. The sworn statement of Morse would be ample answer to such unsupported assertions. This witness does not assail in any way the correctness of the account of receipts as filed by Morse. He merely says if certain things had been done, he thinks certain money might have been realized. As to his having heard statements or affidavits that Mr. Kendall had received $200,000 for services rendered to Morse, the explanation is very simple. If O'Reilly ever did hear anything of that kind, it must have had reference to the amount of stock received by Mr. Kendall. The statement A—shows that the aggregate of stock received by Morse, at its nominal value, would have been over $500,000, while its real value amounts only to $151,795—much of that stock being actually valueless. (See statement A of account and the testimony of T. P. J. Shaffner.) In this view Mr. O'Reilly's allegation may have some ground of truth, but in no other light. And we desire here to call especial attention to the immense difference between the nominal remuneration in stocks that Morse has received and his actual receipts. Much of the stock received never had and never will have any actual value. He was forced to take this stock in payment for his patent rights. This view will also explain a false statement which the press and the opponents of this extension have industriously circulated, viz: that Morse has realized an enormous fortune from his invention. He is, indeed, rich in worthless telegraph stocks. If all these stocks had been worth their nominal value, he would not be driven by his necessities to ask for an extension. It is in consequence of the worthlessness of the medium in which he was paid, that his actual remuneration has been so small. This witness (O'Reilly) was called at the last moment to prove a point, of which there had been no previous notice, and which could not possibly have been anticipated by us. Yet we procured such rebutting testimony as the shortness of time would permit. George Wood and

William Stickney contradict and expose this witness most completely.

Both of these witnesses had access to Professor Morses' books, papers, correspondence, and accounts, and were acquainted with his transactions with O'Reilly. They prove that the suits by Morse *vs.* O'Reilly, were instituted and carried on by the advice of eminent counsel and were deemed essential to protect the interest of Professor Morse; that learned counsel believed H. O'Reilly to have violated his contract with Morse and to be acting in bad faith, and that the connection of O'Reilly with Morse was (in the language of Wood,) "one of long continued wrongs inflicted by O'Reilly, and losses sustained by Professor Morse, and it is my conviction that throughout the entire agency of said O'Reilly, he has sought with strange perversity and enmity to sacrifice the interest and reputation of Professor Morse, when a contrary course could not but have been as a mere matter of policy to his advantage.'"

That further continuance of this connection, so far from being profitable to Morse, would in the language of ; Wood, have been the reverse." "Professor Morse has derived from Henry O'Reilly, no pecuniary benefit nor any other benefit, but on the contrary nothing but losses and wrongs inflicted with singular avidity and untiring zeal."

Mr. Wood is borne out in this opinion fully by that of Mr. Stickney. To the vague interested and contradictory assertions of O'Reilly, upon matters of opinion, we submit that the testimony of these disinterested and intelligent witnesses with full opportunity of observation and judgment, is a satisfactory reply.

There remains to be considered but one question; has Professor Morse obtained a reasonable remuneration for the time, ingenuity, and expense bestowed on his invention and the introduction thereof into use? The account of receipts filed by Professor Morse, embraces four items he has received.

1st. Cash for direct sales of patent rights to the amount of $36,630 48.

Secondly. He has received stocks in payment for patent rights, a part of which he has sold for $21,200, and

Thirdly. He has received as dividends on stocks, the sum of $17,000 75.

Fourthly. He has in hand stocks unsold which, In case the patent be not extended, will be worth not more than $85,950 00.

In reference to these items, it is only necessary to remark by way of explanation upon the 3rd and 4th items. The third item embraces dividends amounting to $17,000. Some doubt was entertained whether this item properly belonged to the account; but it was put in because Professor Morse chose not to expose himself to any imputation of concealment. If he had received money instead of those stocks, it would not have been maintained by any one, that he is required by law to render an account of the interest he might have received on such sums, made by the use of that money. Yet, there is little difference between an income from stocks taken in payment for a patent right, and the interest in money received on the same account.

It has never been required of a patentee to add together the moneys which he has from time to time received for his invention and charge himself also with the aggregate interest thereon down to the date of his application for an extension.

Thus, suppose the aggregate cash receipts of a patentee had been $50,000, during the first seven years of his patent term, he would not be required in his statement of account to add thereto the sum of $21,000, as interest. We doubt if such an estimate was ever given by a patentee on an application for an extension, and upon the same ground we submit that Morse, having charged himself with the value of stock received by him he should not be required to charge himself also with the dividends thereon.

The amount of dividends received, have however been included in Morses' receipts and are with this explanation submitted for the Commissioner's inspection in the case.

The fourth item of receipts embraces the estimated value of stocks on hand, and this value has been estimated upon the hypothesis of the non-extension of the patent.

The value of these stocks depends upon the amount of present and future dividends which will probably be received thereon by the stockholders. The expense of conducting the business involves the cost of maintaining lines, batteries, operators, officers, &c., and must remain about the same, no matter what amount of business may be done.

It is also well known that there are various rival telegraph systems lately introduced, which have been held to be infringements of Professor Morse's patent, and which have been from time to time started in various parts of the country. It is shown in testimony (see Shaffner's evidence) that with the experience now possessed by telegraphic constructors, and the increased facilities for supplying wires, posts, insulators, &c., new lines of telegraph may now be constructed at a cost vastly less than that at which existing lines were constructed. Should this patent not be extended, capitalists perceiving these facts and no longer deterred by the fear of infringing, would readily be induced to unite with the agents of various modifications of Morse's telegraph, and even to use his identical instruments upon newly built lines. What would be the practical result? The new companies having been at no expense for patent rights, and having constructed their lines at comparatively small costs, could afford to transact their business at a rate much less than the old companies. The new companies would be induced to work at lower rates with the express view to draw to themselves business from the old lines. The old companies would then be forced to reduce their charges to below what would yield dividends, or to suffer their business to pass into the hands of others. Either course would be ruinous, and would rapidly depreciate the value of stock in the old companies. The extent of this depreciation cannot be stated with certainty; but it is believed that it could not on any lines be less than 50 per cent. The experience of the past has shown that in many cases, the effect of rival lines has been far more disastrous, and in some cases rendered profitable lines intirely valueless.

The attention of the Commissioner is referred on this point to the account filed in this case, setting forth the numerous lines rendered profitless by opposing and infringing lines, and to the testimony of T. P. Shaffner, who is well qualified to judge on this question from his long experience in telegraph matters. Shaffner says:

"Owing to the rivalry of parties claiming rights under conflicting systems of telegraphing, the utmost difficulty has been the experience of all in procuring subscriptions for stock in building lines, and thus the patentees of the Morse telegraph have lost many thousands of dollars.

"On the St. Louis and Missouri River Telegraph line, Morse was prevented from receiving more than one-half of the value of his patent. Nearly four thousand dollars would have been paid him if it had not been for the competition placed in the way by rival and conflicting systems as before mentioned.

"To what extent Professor Morse has been deprived of receiving consideration or compensation for the patent right by the efforts of rival parties under conflicting systems, some of which the United States Courts have decreed as infringements, is difficult to ascertain, but including the losses sustained by the nonpayment for his right, by their use of his patent at defiance, the sum will exceed a million of dollars. This opinion is formed from my experience and observation for the last seven years.

"I am probably better acquainted with the revenue of the respective telegraph lines of America than any other person, and I am confident and thoroughly convinced by many estimates that the value of the stocks in the existing lines on the main routes throughout the country will beyond all doubt be reduced in value at least seventy-five per cent.; that is, the stock will not be worth more than one-fourth as much as their present value, if the said Morse's patent of 1840 is not extended.

"I have been directly or indirectly connected with the New Orleans and Ohio Telegraph Company since its first commencement, December, 1847, and am well acquainted with its financial condition and the value of its stock.

"Owing to the efforts of rival parties and the extension of rival lines, and by the use of other systems of Telegraphing (since decreed by the United States courts as infringements,) since the establishment of the first section of the line, the company has been hindered from making any money for dividends, or enough to pay expenses, until the past year, causing the line to run in debt many thousands of dollars. This competition also prevented the constructors of the line from procuring money to erect the line.

"This company is largely in debt, and, if the Morse patent is not extended, the value of the stock will be worthless, and the line will have to be sold for its debts at a total loss to the shareholders.

"The non-extension will open the route for a new line which can be built upon a reduced capital, and with the advantages of experience dearly purchased by this line, successfully compete with the old company and totally destroy its existence.

"As to the St. Louis and New Orleans Telegraph lines I will state, that I was the builder of it, and was for nearly three years President of the company.

"The line is now in debt, and if the patent is not extended, a new line can be built avoiding many of the difficulties ascertained to exist by the experience of the present line, and thus totally destroy the property of the company, which would be a total loss to the stockholders in the said St. Louis and New Orleans Company."

The account states that there are certain unliquidated claims against F. O. J. Smith, and that a suit was commenced in New York, against said Smith, to recover this amount; that said Smith denies in his answer, under oath, all indebtedness to Morse. This matter has been once submitted to the proper tribunal in New York, and after full argument, was decided adversely to Professor Morse's claim. It is submitted that this claim belongs to the class of unliquidated claims against Infringers. And that a patentee having failed, after due prosecution of his suit to recover, he is not to be charged with it as receipts. The circumstances under which the claims arose, are, however, fully set out in the applicant's accounts.

The account, also, mentions thirteen lines of telegraph, in which payment has been made in stock to the patentee; but that said stock has at present a very reduced value, and that this value depends almost entirely on the result of the present application for extension.

On this point, also, we would adduce the testimony of T. P. Shaffner, before cited. The four items of receipts contained in Morse's account, after allowing a contingent estimate of fifty per cent. for reduction on stock now in hand, in case the patent be not extended, are as follows:

For cash received,	-	-	$107 021
Stock in hand and unsold,	-	-	85 050
Total receipts,	-	-	$193 871

From this is to be deducted Morse's expenditures, which are set forth at length in his statement, and embrace:

1. Cash paid for expenses arising under patent, and specially set out in paper C of account, - - $1,083 19

2. Amount paid in litigation against infringers, as per
statement D - - - - - - - $6,037 68
3. In litigation with F. O. J. Smith, as per statement E, 3,757 39
4. In repurchase of Dr. Gale's interest, as per statement F, - - - - - - - 15,000 00
5. Expenses incurred in originating and developing
invention, as per statement G, - - - 74,518 00

Making a total of expenses of - - $102,806 26

On the foregoing items of expenditure, it is perhaps unnecessary to make any observation, except by way of explanation on the last.

The four first are such as should be legitimately and clearly included within the account, and the fifth has by the practice of the Patent Office been declared to be an equally legitimate and proper item of charge.

Professor Morse devoted himself for twenty-one years to the study of portrait painting, and at a cost of not less than $1,000 per annum, making $21,000. After acquiring his profession, he was compelled, in 1838, to abandon it, in order to devote himself to the introduction of his telegraph. At the annual value of of $3000, which he fixes on his profession, this would involve an aggregate loss of $48,000, down to the year 1854. Professor Morse, if cut off from future receipts from his telegraph patent, cannot look to his profession for remuneration; for, as is well known, a profession abandoned for sixteen years cannot be readily recalled. For the remainder of his life he will be deprived of revenue from that source. Hence, he is justified in considering the capital invested in acquiring that profession as an entire loss, and as a loss additional to that which he would have derived from the daily pursuit of his profession during the sixteen years of its abandonment. He must lose the future, as he has lost the past interest in his investment, and hence he is right to charge for both the original principal and the past interest. The amount of expenditures being then correctly stated, there appears a balance of receipts by Professor Morse, from all his telegraphic inventions, of $90,874.

Professor Morse was the inventor of an additional improvement besides that patented in 1840. This is commonly called his local circuit improvement. This invention is one of great value to telegraph lines, and is secured to Professor Morse by a patent originally granted April 11, 1846. This improvement was introduced and patented simultaneously with the establishment of the New York and Washington line, the first line of telegraph, and all the Morse telegraph lines of this country have used it from that time to the present day. The right to use this improvement has been included in every grant made subsequently to April 11, 1846, and was conveyed under the term, "all future improvements," which had been inserted in grants and licenses under the original patent. The novelty and, value of the improvement embraced, in the second patent of Morse, is recognised by Judges Grier and Kane in the the case of French vs. Rogers, before cited, and by the United States Supreme Court, in the case of O'Reilly vs. Morse, page 17. This improvement being of confessedly great practical value, and having been universally employed in connection with Morse's original invention, it is fair to consider that one-half of the amount of profitable receipts should be credited to the second patent. This would make the amount received by Prof. Morse, from his first patent, $15,437 48. Is that sum a reasonable remuneration for Morse to receive from the American public in view of the grandeur of this invention; of his perseverance; of his sacrifice and his toils; in view of the immense benefits derived from it; in view of the magnificent telegraphic system of America, of which he was the founder; in view of the merit awarded to him by the highest judicial tribunal in the land—of being, in the words of Chief Justice Taney, the first and "original inventor of the telegraph." The amount received by him is no greater than was recently deemed by the Commissioner an inadequate remuneration to the gentleman who placed side discs on a fan blower.

Compare Morse's receipts with the value fixed by Mr. Tucker, President of the Philadelphia and Reading Railroad Company. He says it has saved $10,000 on his road in a single day; that the railroad is 91 miles long, and, he has no doubt, the company saves more than $25,000 a year by the use of this telegraph, and that it is proportionably valuable to other roads. In the official abstract of the Seventh Census, page 101, it is stated that there were 13,266 miles of railroad in operation in the beginning of 1853, and 12,031 miles in the actual course of construction, making a probable total length of railroad completed by this time, or about to be completed, of 25,917 miles, or 285 times the length of the Reading railroad. If the value to the Reading road be not less than $25,000, then, on the same proportion, it would be worth to the railroad interests of the country six millions and a half of dollars annually.

Mr. Addick says, he thinks one or two hundred thousand dollars would be a small estimate of its value to Philadelphia. It must be much greater to New York; very nearly, if not equally, as great to Boston, Baltimore, New Orleans, Cincinnati, Pittsburg, and St. Louis; so that, even on that low estimate, its commercial value annually would be millions.

In a single case of police arrest at Philadelphia, $15,000 was recovered through its agency only.

Some witnesses say it increases the aggregate business capital of the entire community, 10 per cent.

The citizens of New York, Philadelphia, Baltimore, and Washington voluntarily paid $103,232 last year for its use between those cities only.

The report of the decision of the Privy Council in 1850, on the application for an extension of Cooke & Wheatstone's patent shows that the electric telegraph company paid for the patents alone the sum of 167,688 pounds sterling—nearly \$838,000. (See report published in the London Mechanics' Magazine, vol. 151. page 131.) Our Supreme Court, as before stated, have declared that Morse was a prior inventor to Cooke & Wheatstone; and we have shown that the lines of telegraph in England were only 4,000 miles long, when Morse's lines in this country had reached to over 23,000 miles.

Again, it is to be borne in mind that Morse was forced to part with one-half of his interest under the patent before he could obtain means to introduce it; and it was not until 1847, when the term of 14 years was half expired, that he realized any income from the remaining part. So that he has, in fact, lost three-fourths of the value of the entire exclusive grant originally secured to him by the Government. Necessity, misfortune, and public prejudice having combined to deprive him of three-fourths of his exclusive grant, is it not just and right that the Government should, by its proper officers, extend this patent?

As before stated, the amount of remuneration which Morse will have received in case the patent be not extended, will be \$45,437.

Should the Commissioner *having a due regard to public policy,* extend his patent? Are not the public largely indebted to Morse? Morse not only sacrificed his profession and devoted his time, ingenuity, and money, to the perfection and introduction of his invention, but by his exertions he procured the investment of large amounts of capital. Most inventions after having been perfected and publicly introduced, are at once taken advantage of by the community. The cause of this is plain: Inventions generally consist in improvements on previous machines or processes; persons using these machines or processes, at once perceive the advantage of such improvements, and of their own accord adopt them. Thus, the carpenter whose boards had been planed for twenty-five dollars per thousand feet, immediately perceived the value of a machine which would plane them for four dollars per thousand feet, and adopted it. But the telegraph was a new art, a new institution as it were. It did something which had not been previously done at all, which men had not desired or wanted to do. Hence the grave question arose whether this beautiful, ingenious, wonderful invention was practically valuable. It did not telegraph better than men had telegraphed before because it may be said they had not telegraphed at all. *Cui bono?* was the universal question. The Postmaster General, Cave Johnson, whose position ought to have enabled him to form as correct an estimate as any one, said 'It cannot possibly yield a nett profit." After the perfection and public exhibition of this invention by Morse, there yet remained another great work to be performed. Men had to be taught to use the invention; they had to be instructed to do that by telegraph which they had before done by mail; capitalists had to be induced to expend their money in an experiment; in a great experiment. The public are indebted to Morse for thus inducing capitalists to invest their money in this enterprize. By his exertions in this respect 36,000 miles of wire have been erected at a cost of many millions of dollars; the public have derived the benefit of this investment, as the statement of Professor Morse shows, that the stocks on many lines are nearly worthless. Capitalists have lost their money, but the lines of telegraph exist and the public have enjoyed and still enjoy the benefit of them; and we submit that on this view of the case the public are largely indebted to Morse for the amount of capital which he has thus directed into this channel. He is not only the inventor of an ingenius instrument, but he is the founder of a great American system or institution, the American telegraph—an institution which has done more to bind together the various parts of our Union, than any other since the adoption of the Federal Constitution—an institution which has probably fixed the location of our seat of government at Washington, and which will determine largely the future destiny and character of our country.

In view of the question of public policy, in connection with this extension, this application commends itself to the Commissioner on another ground, viz: the manner in which the exclusive grant has been exercised under the patent for the past fourteen years. It has been in no sense an oppressive monopoly.

The rates of charge made for telegraphic services in this country, have been extremely moderate. The official abstract of the Census, contains the following comparative estimate of charges for telegraphic communication in this country and in England.

"The charge for transmission of despatches is much higher than in America, one penny per word being charged for the first fifty miles, and one farthing per mile for any distance beyond one hundred miles. A message of twenty words can be sent a distance of five hundred miles in the United States for one dollar, while in England the same would cost seven dollars."

And it is to be remembered that these small charges have in a great measure diminished the revenue of the patentee. At the same time this experience of the past affords a fair guarranty that, in case the patent be extended, the public will not have reason to complain in any manner of the future exercise of the rights secured thereby.

In this argument no notice has been taken of the high commendation passed upon Professor Morse in foreign countries. The action of the convention of deputies at Vienna, wherein all the German and Austrian delegates resolved to adopt the Morse Telegraph; the State medal granted to Morse by the King of Wurtemburg; and the similar testimonials received from the Grand Sultan of Turkey, the King of Prussia, and the Academy of Industry of France, have not been alluded to; although as indirect evidence of the ingenuity and value of the invention they should be considered, perhaps, of some weight.

Having thus shown the grounds on which we ask for the rejection of arguments filed by opposing parties, having replied to such objections as we could anticipate, and having analysed the testimony on behalf of the applicant, we submit that a proper case has been made out, to justify the Commissioner in extending the letters patent as prayed for.

GEORGE HARDING,
P. H. WATSON,
Of Counsel for S. B. F. Morse.